Marine Life: Adult Coloring Book See Creature Collection I

Copyright © 2016

For resale and distribution information, please contact us via www.booboone.com

Walleye Fish

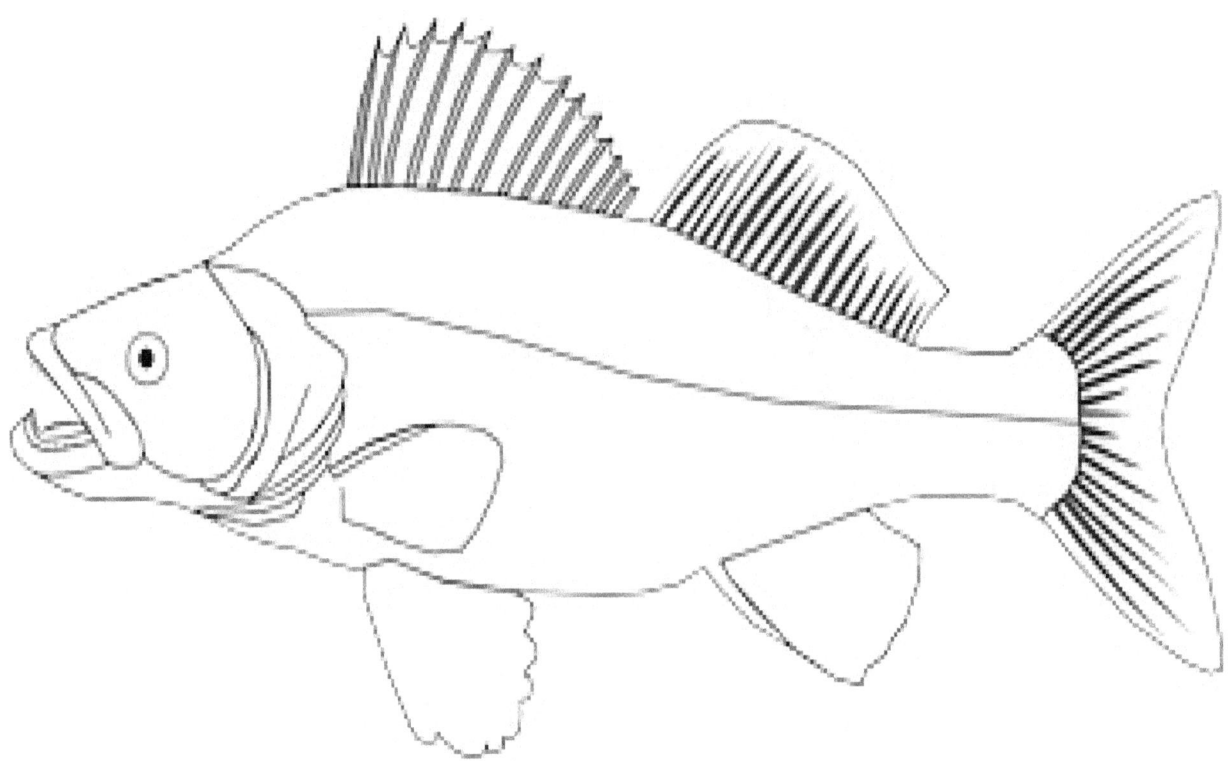

Ophichthus Eel

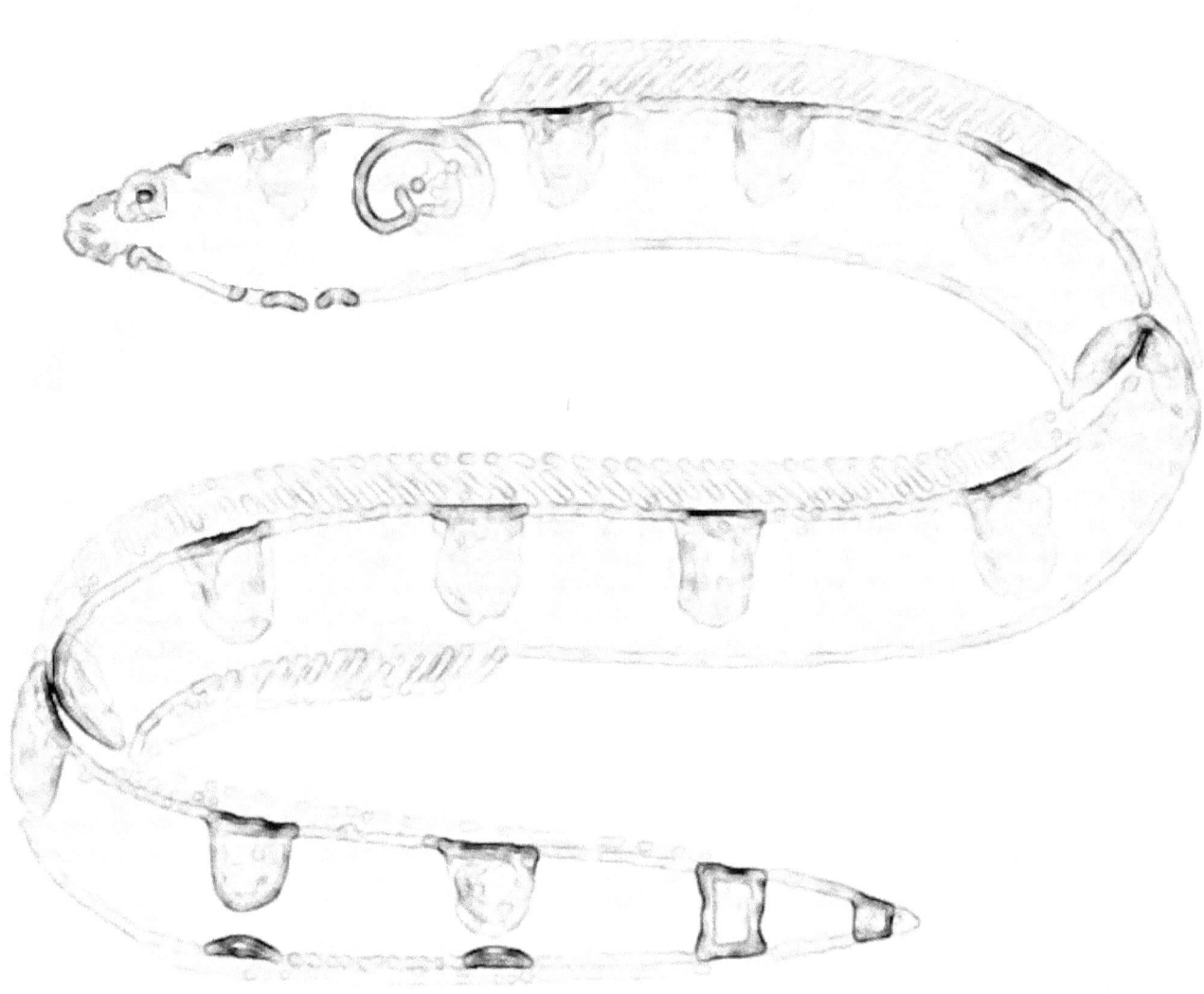

Hammerhead Shark

Whale

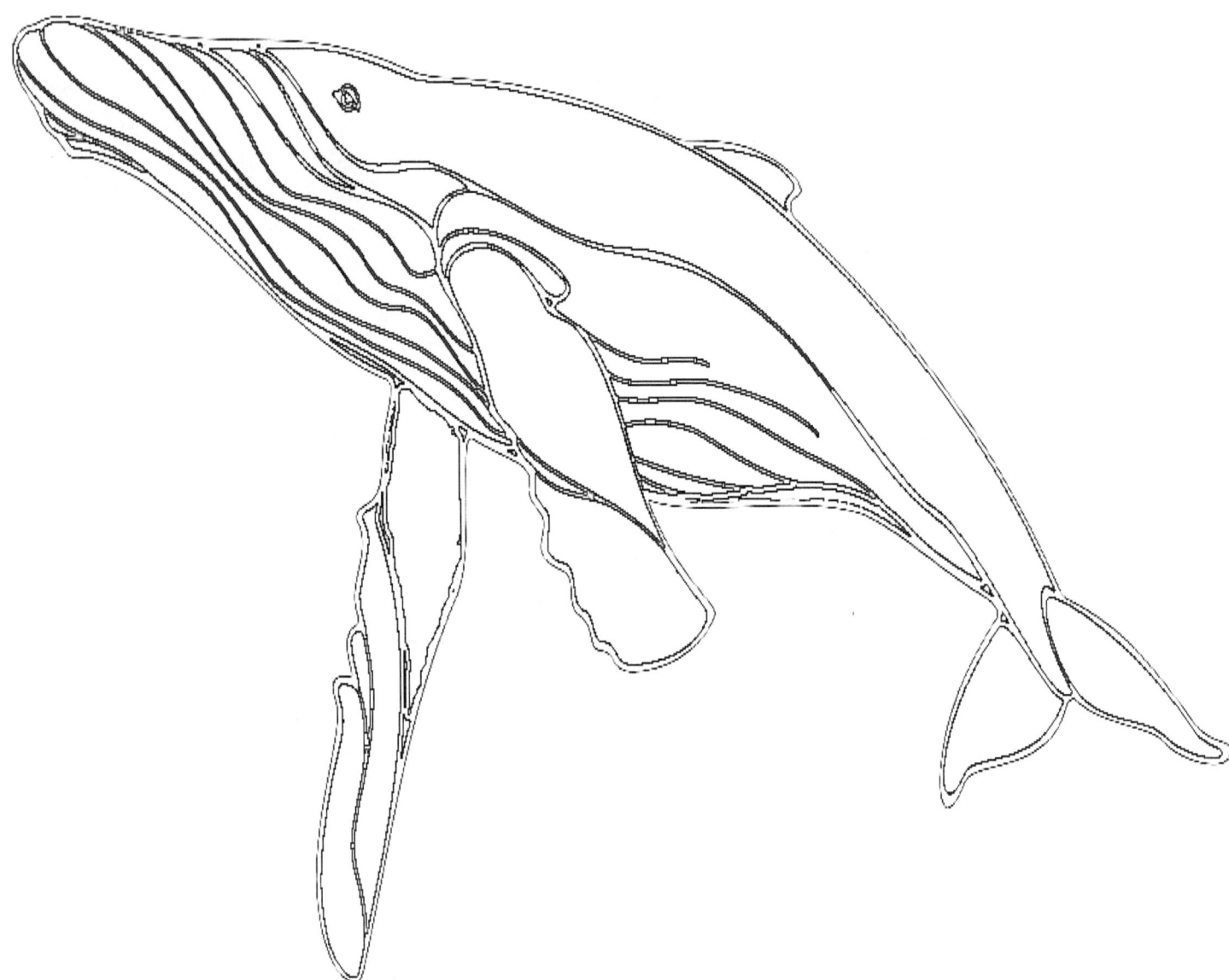

Trout

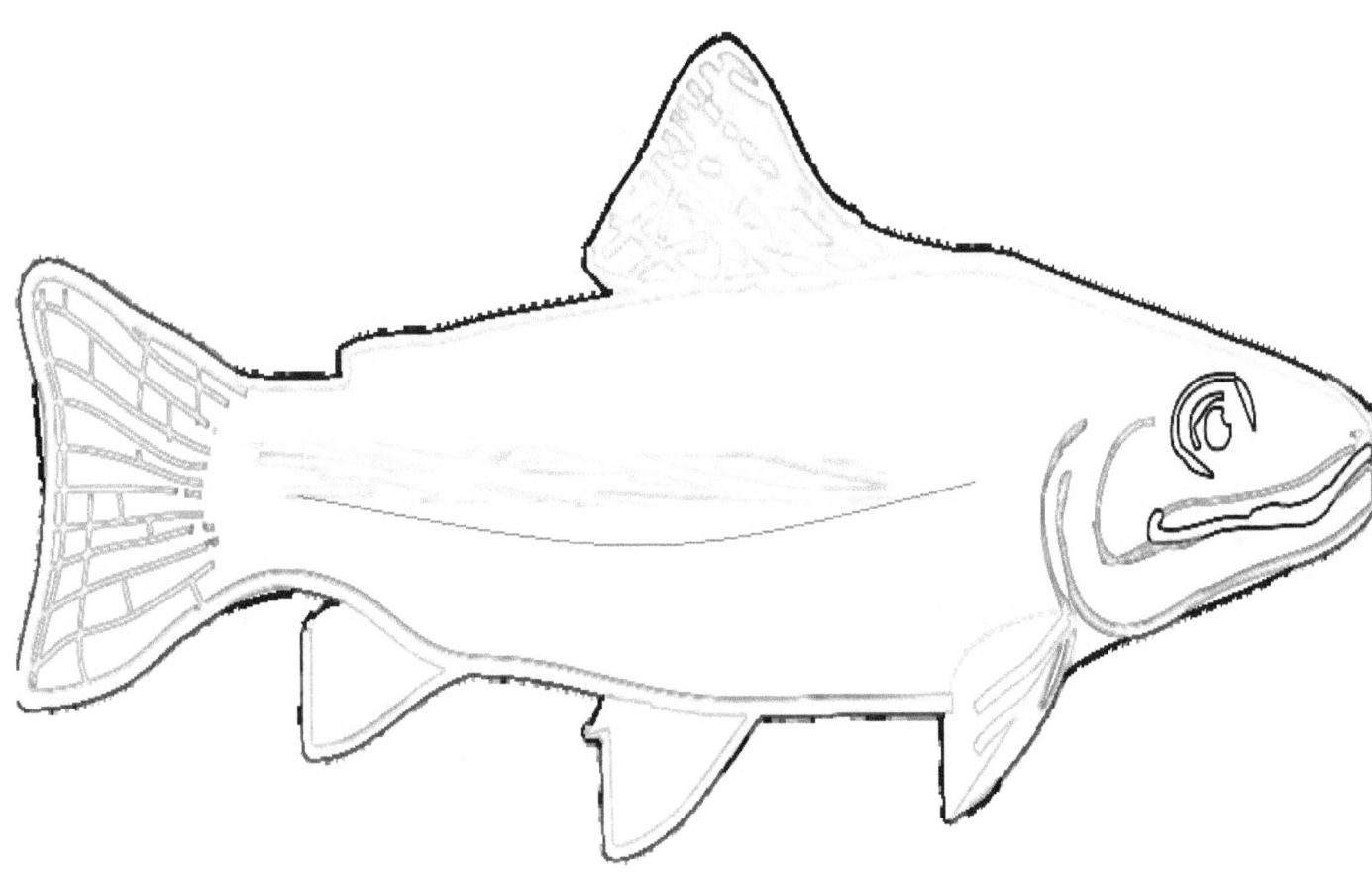

Koi Fish Art

Kokanee Salmon

Leaf scale gulper shark

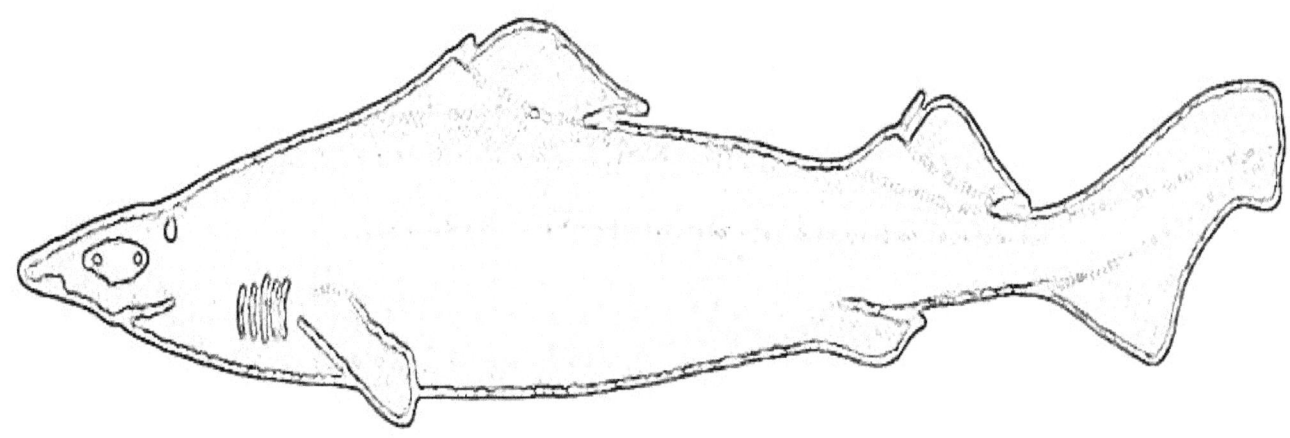

Long-nosed Chimaera

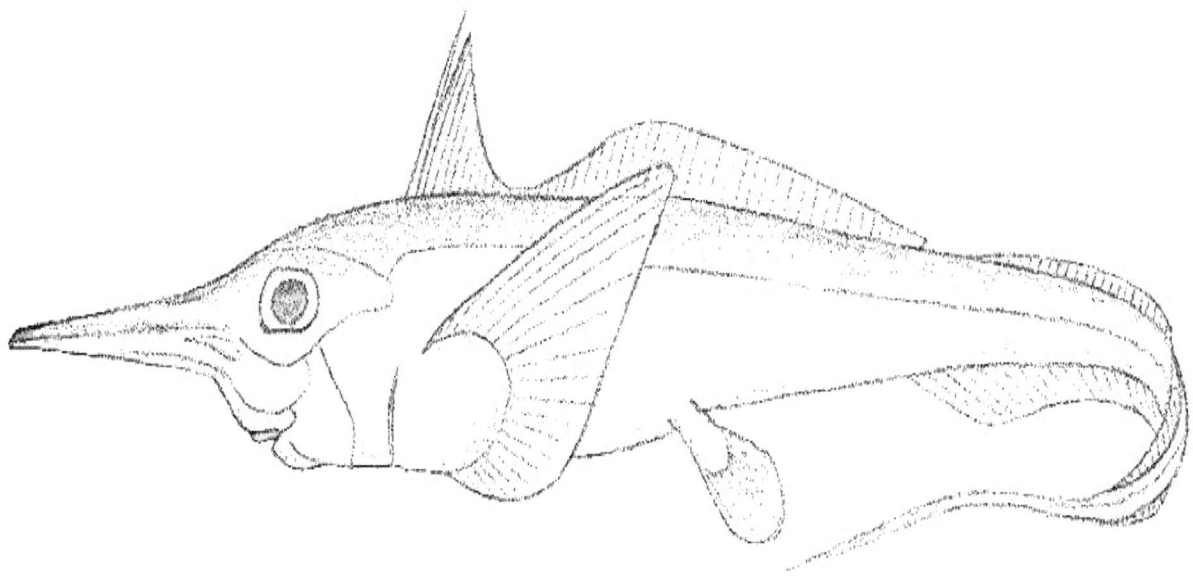

Grey Fish

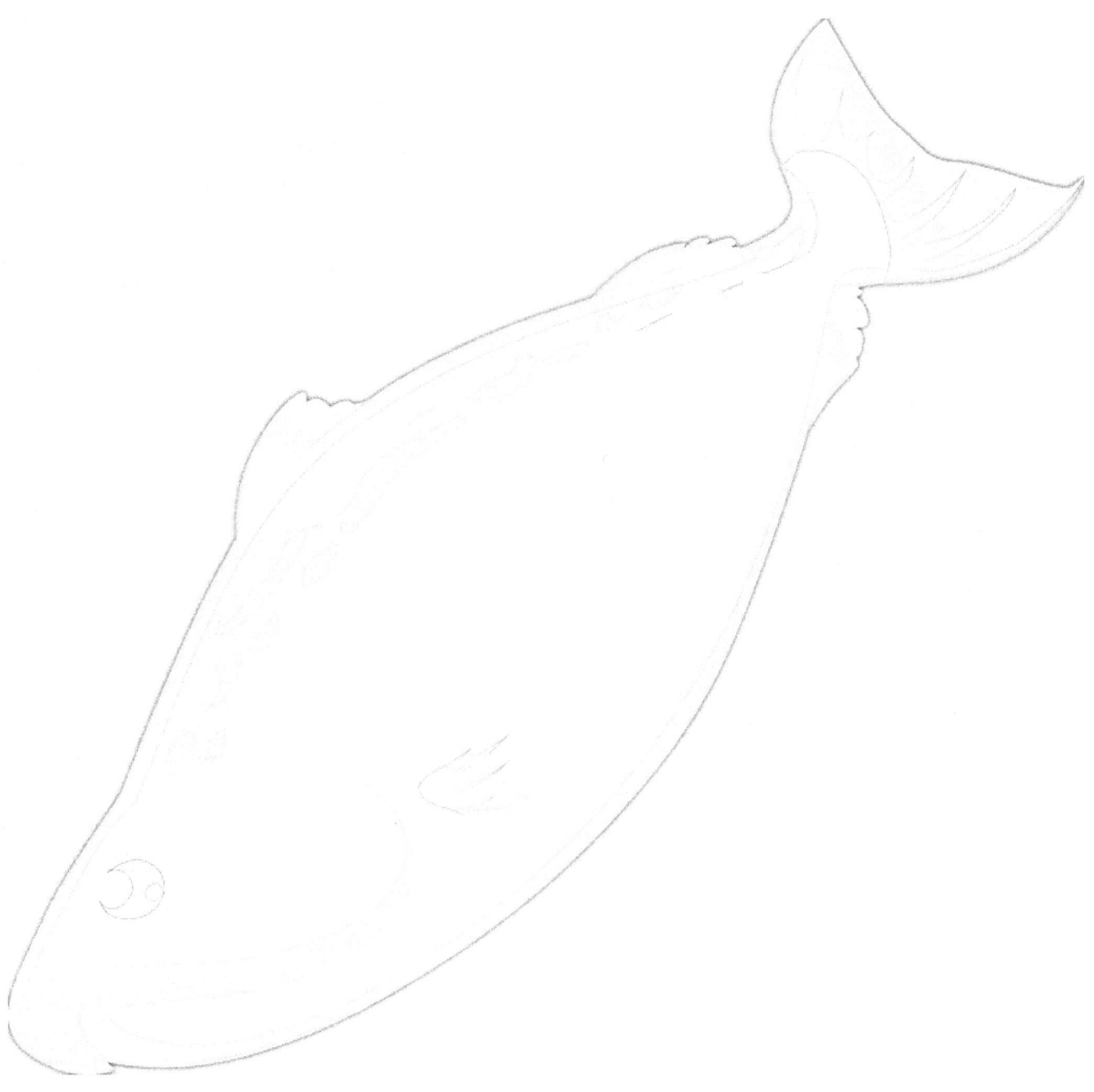

Manatee by Lena London

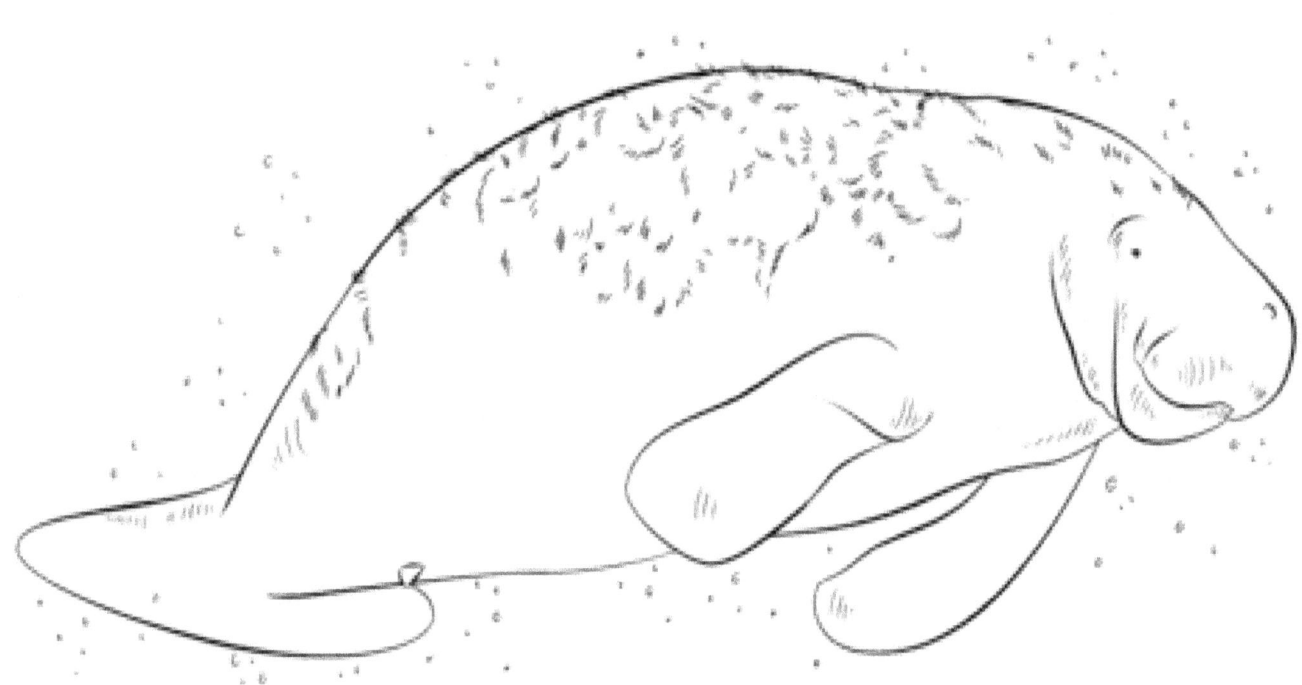

Tropical Fish

Marlin

Molva (Common ling)

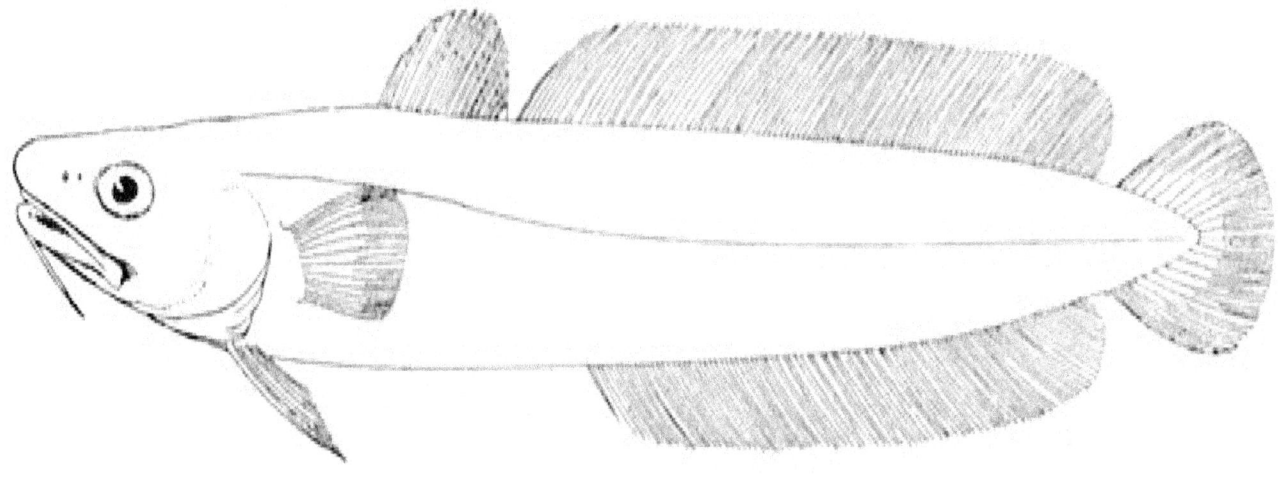

Narwhal

Orange Roughy

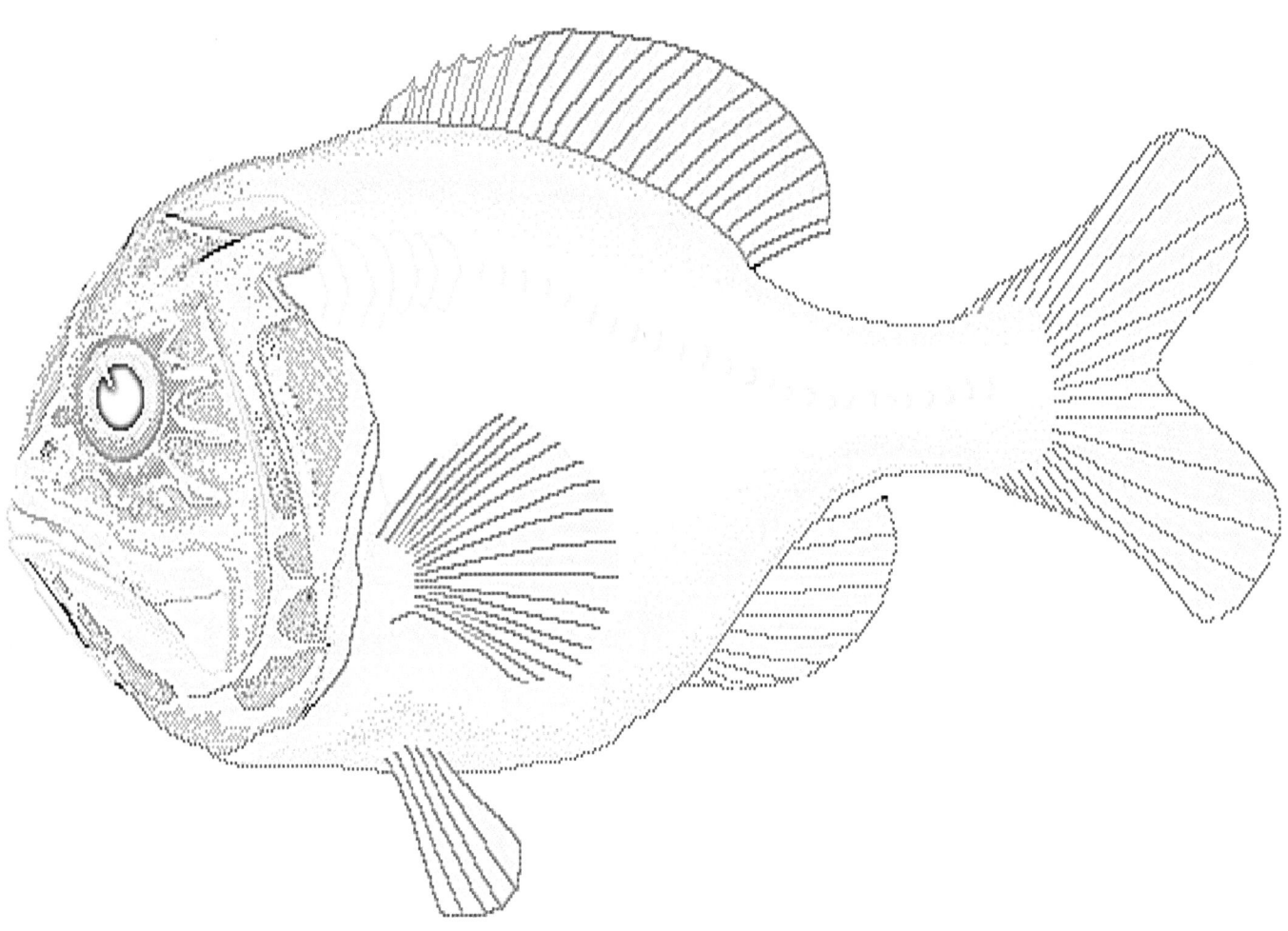

Common Cod

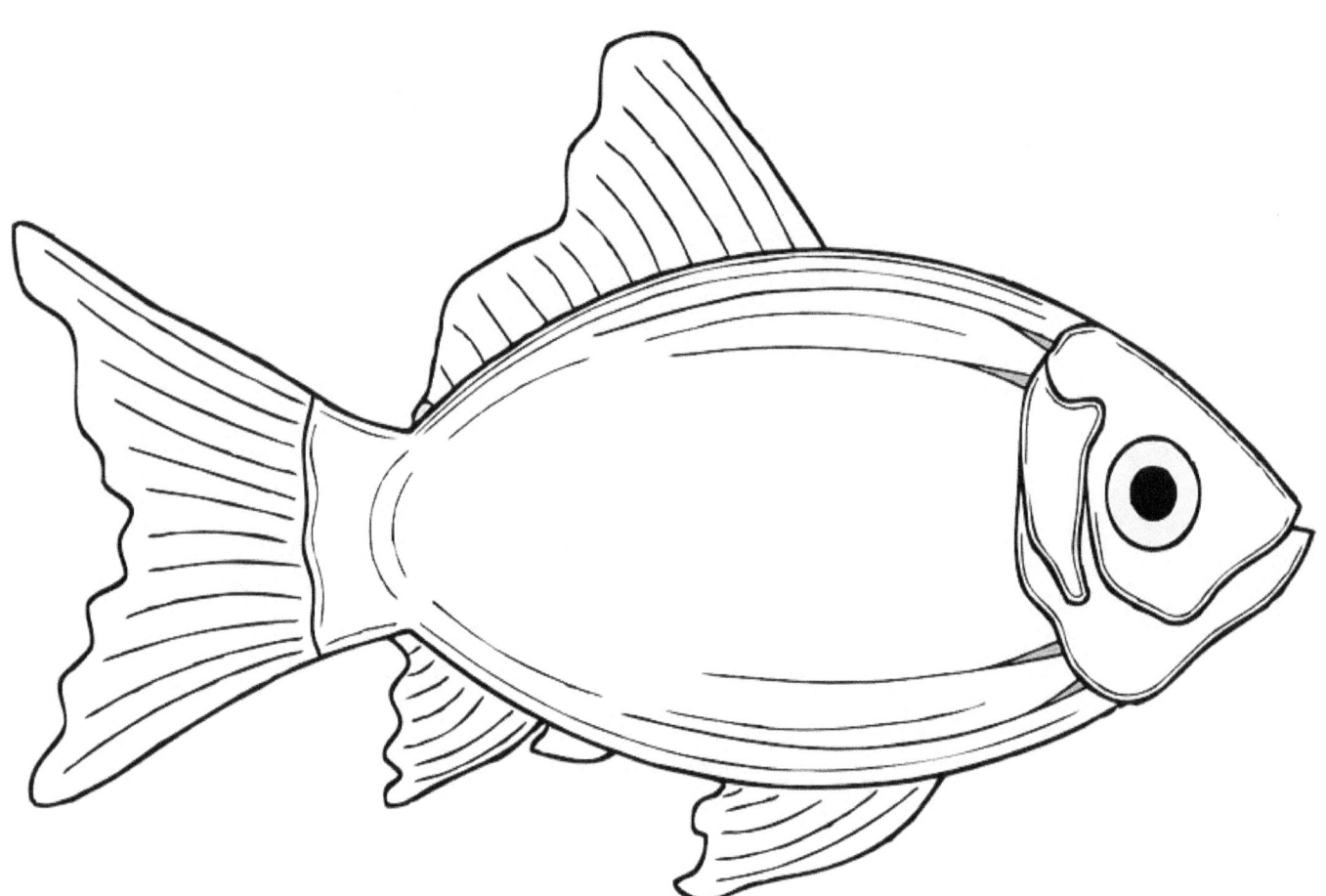

Aquatic Fish

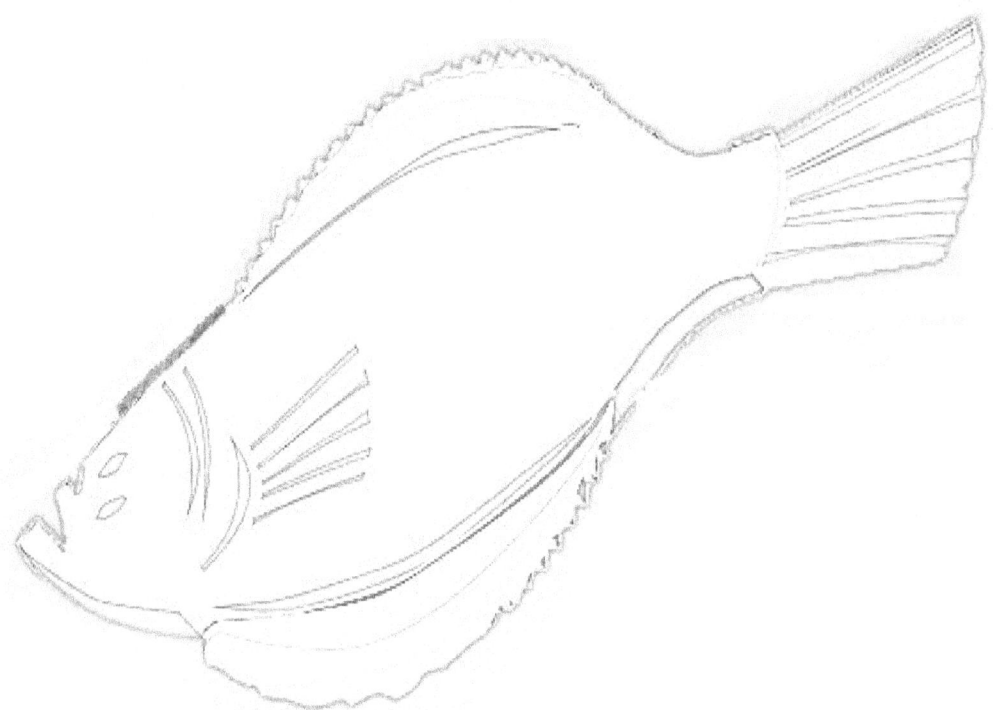

Queen Angelfish

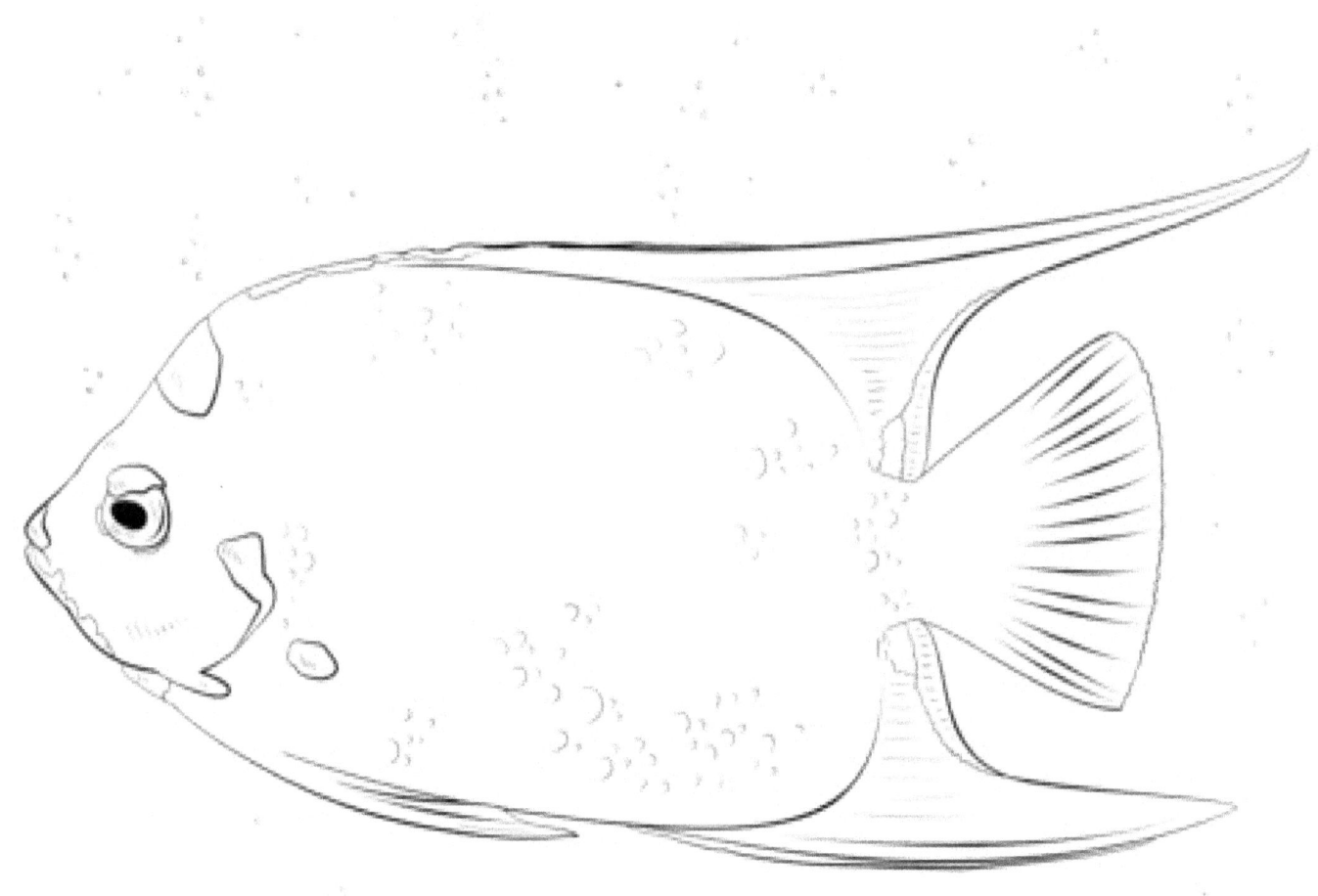

Red-bellied piranha

Southern Right Whale (Eubalaena australis)

Black and Yellow Tropical fish

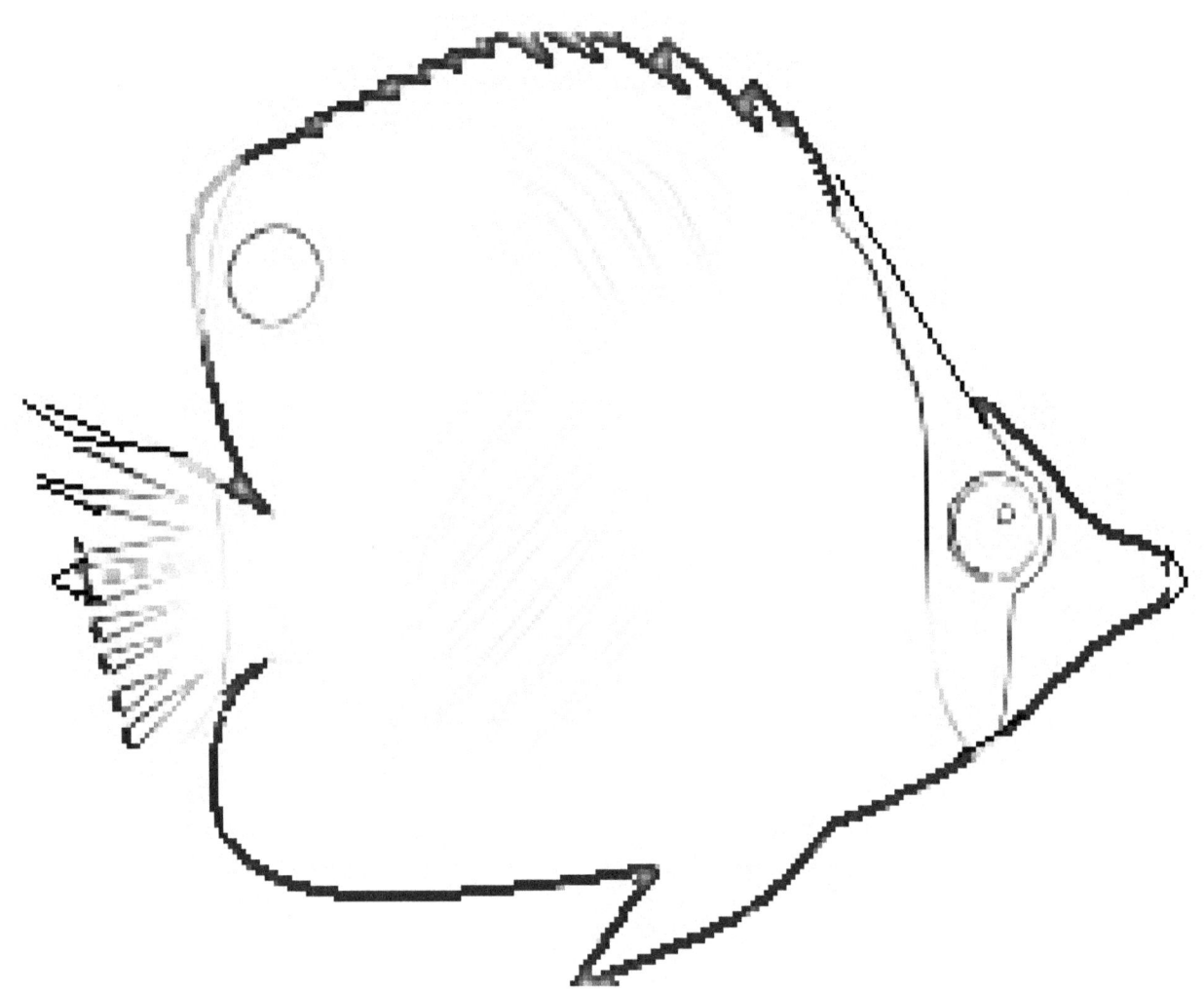

Surgeonfish

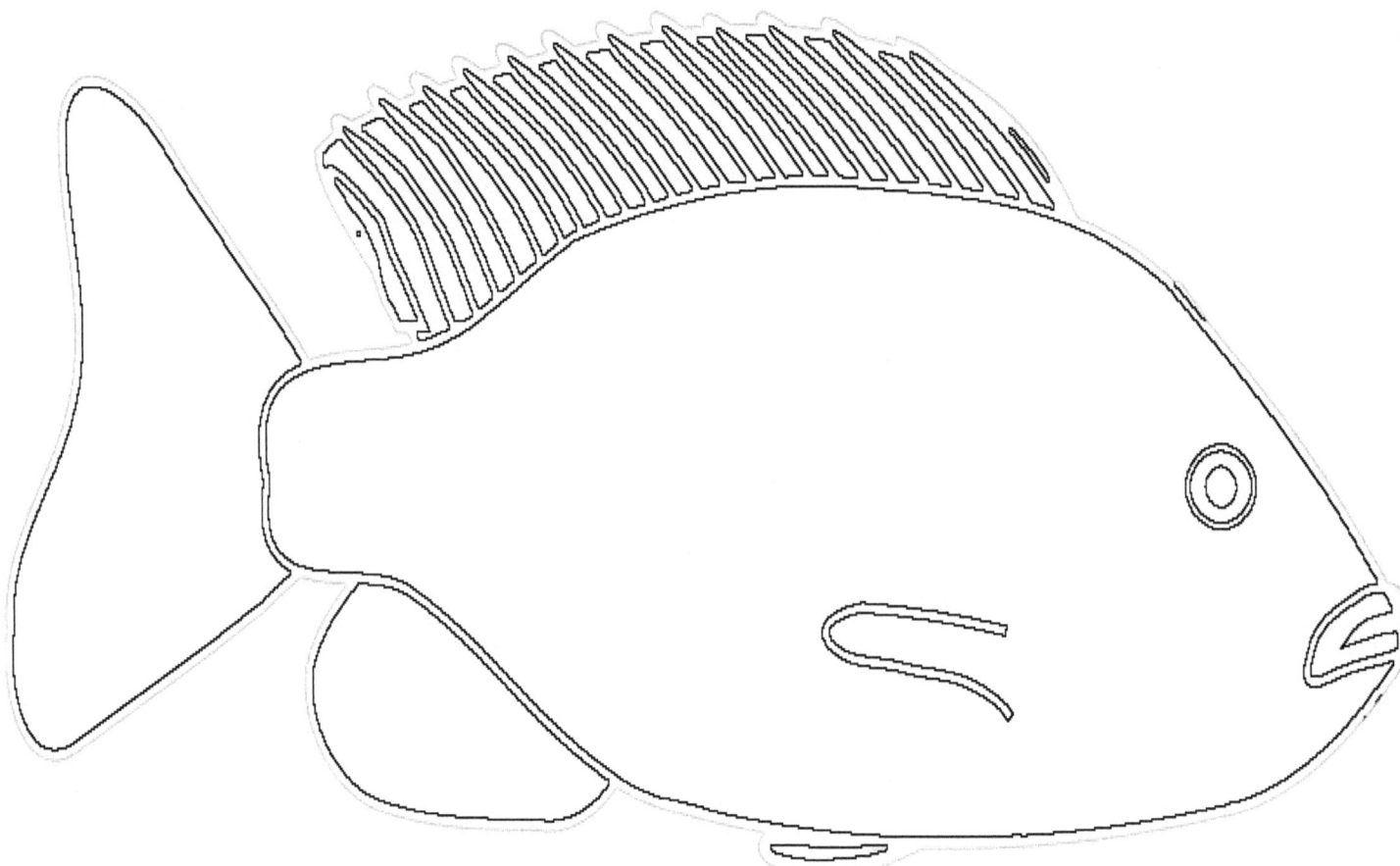

Flying Fish

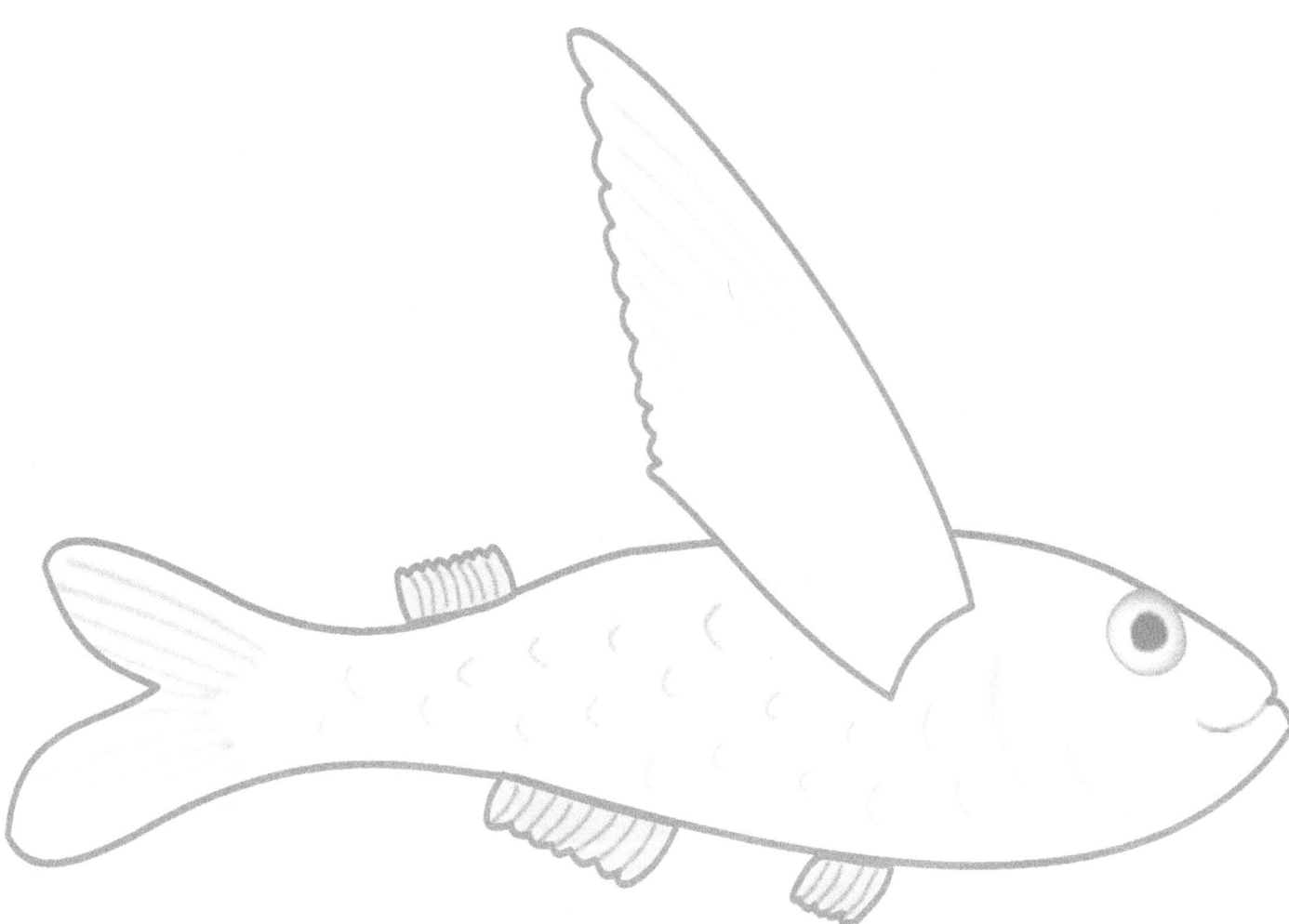

Salmon

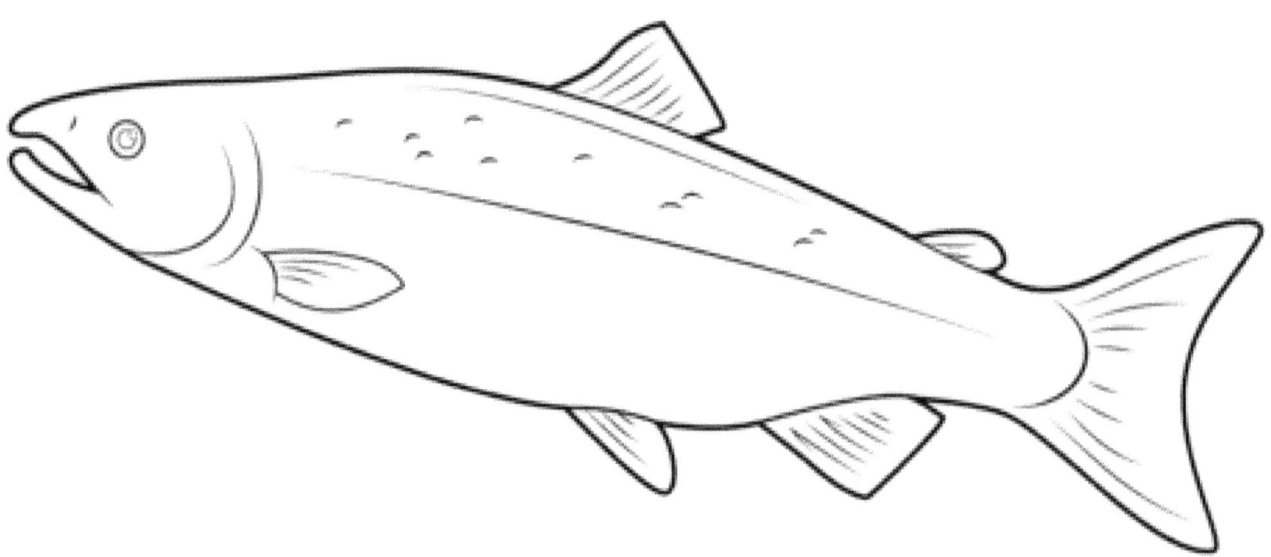

Sea Lion

Lemon Shark

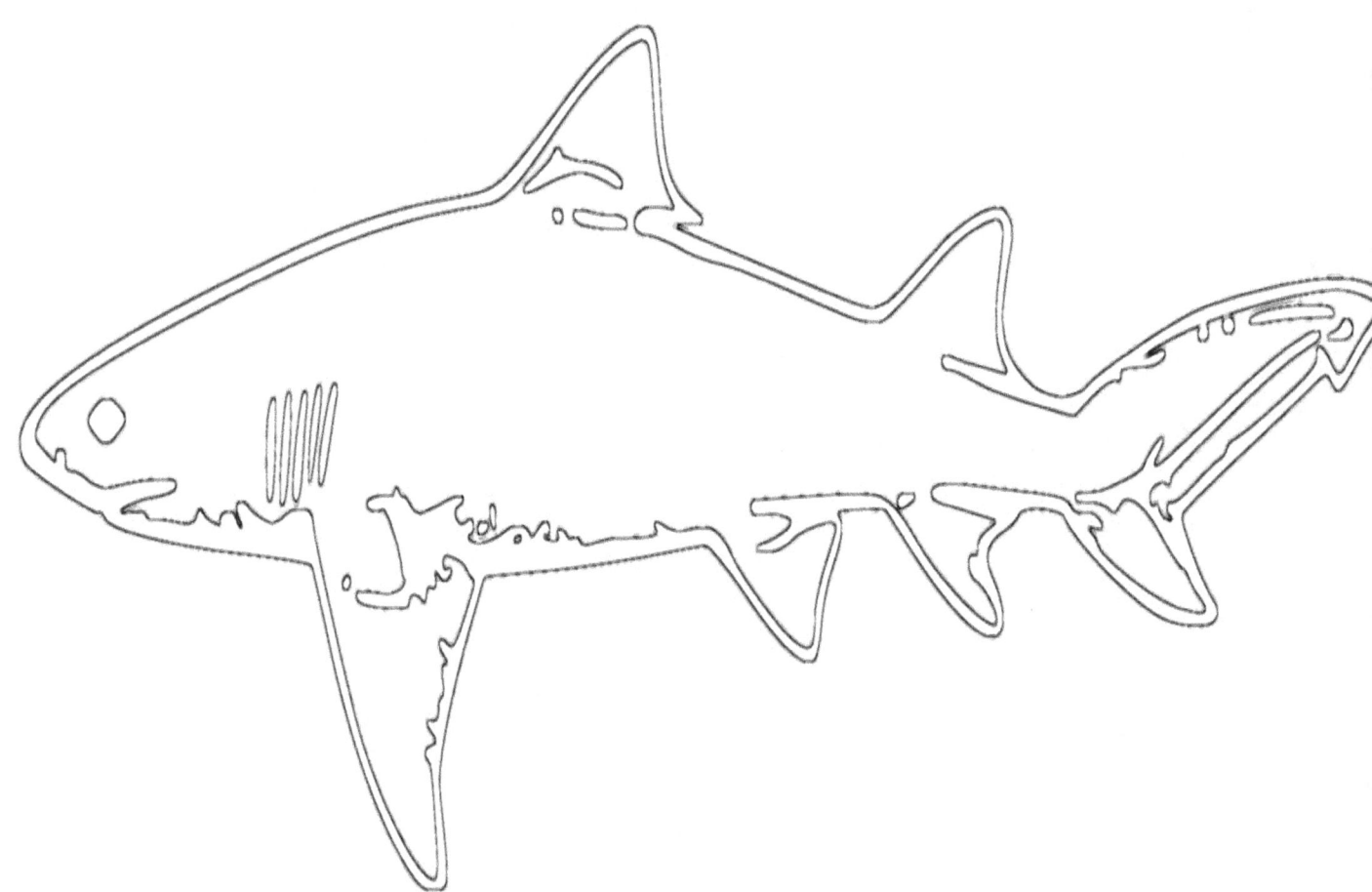

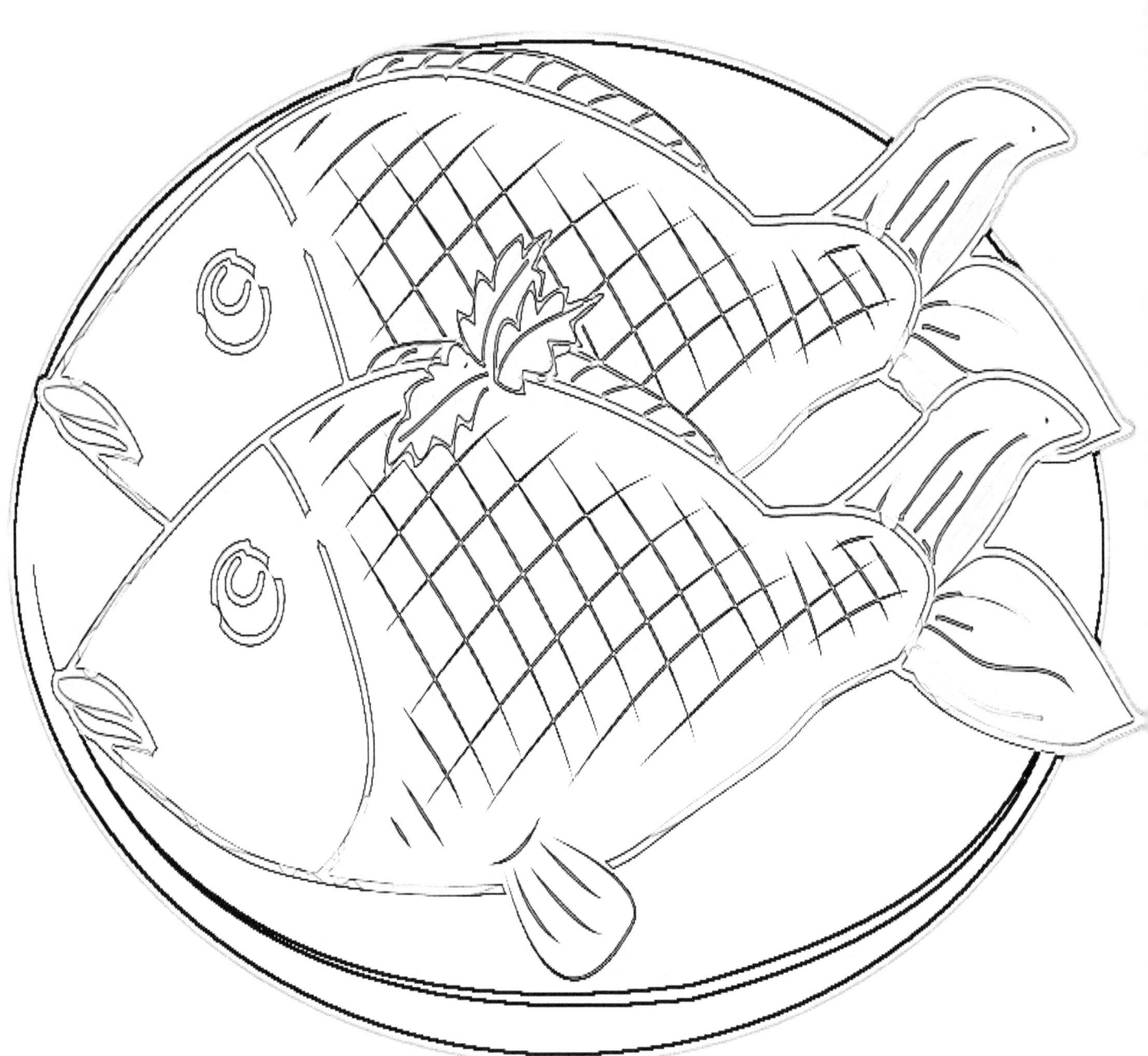

Seahorse

Shrimp

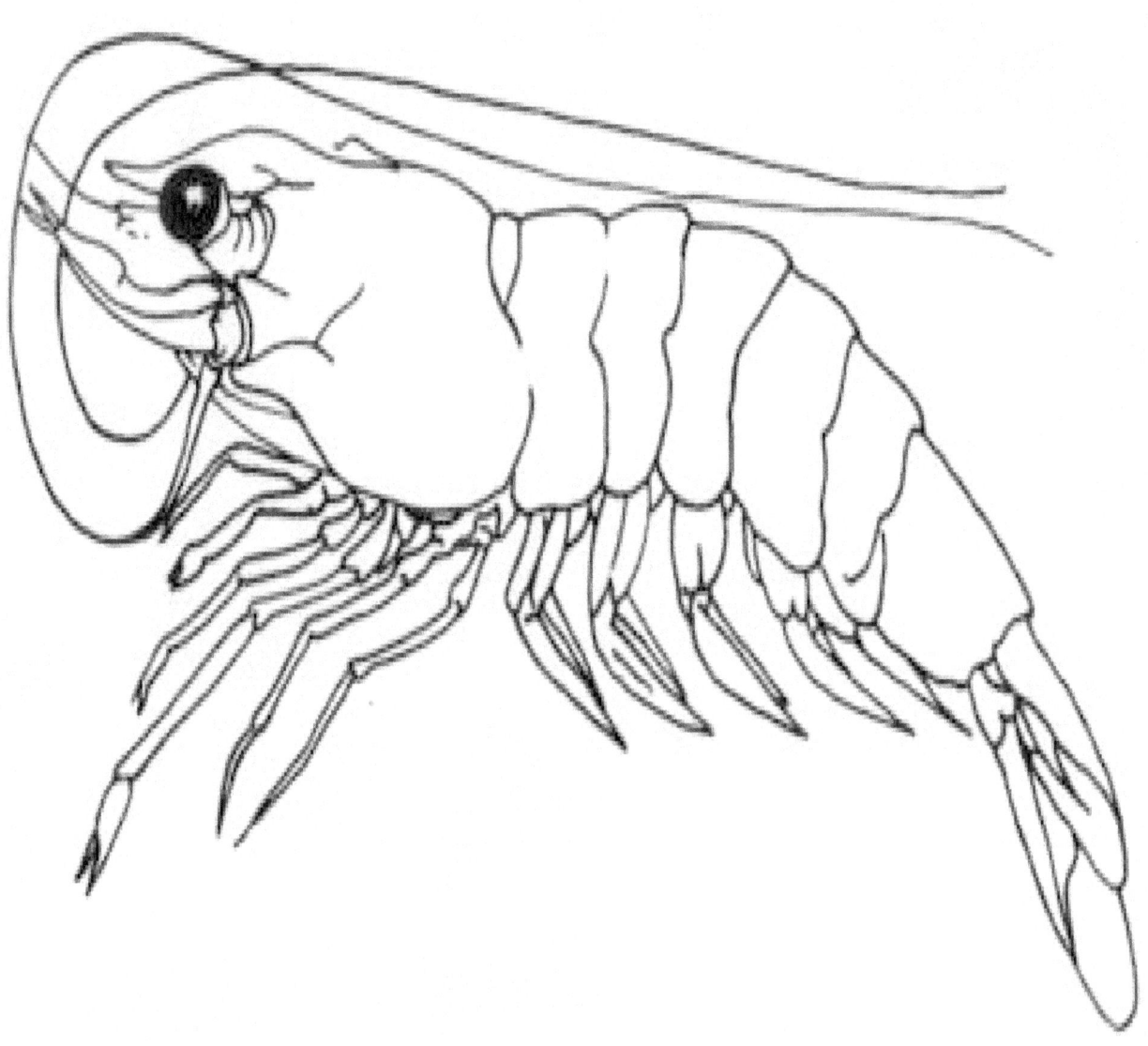

Sockeye Salmon

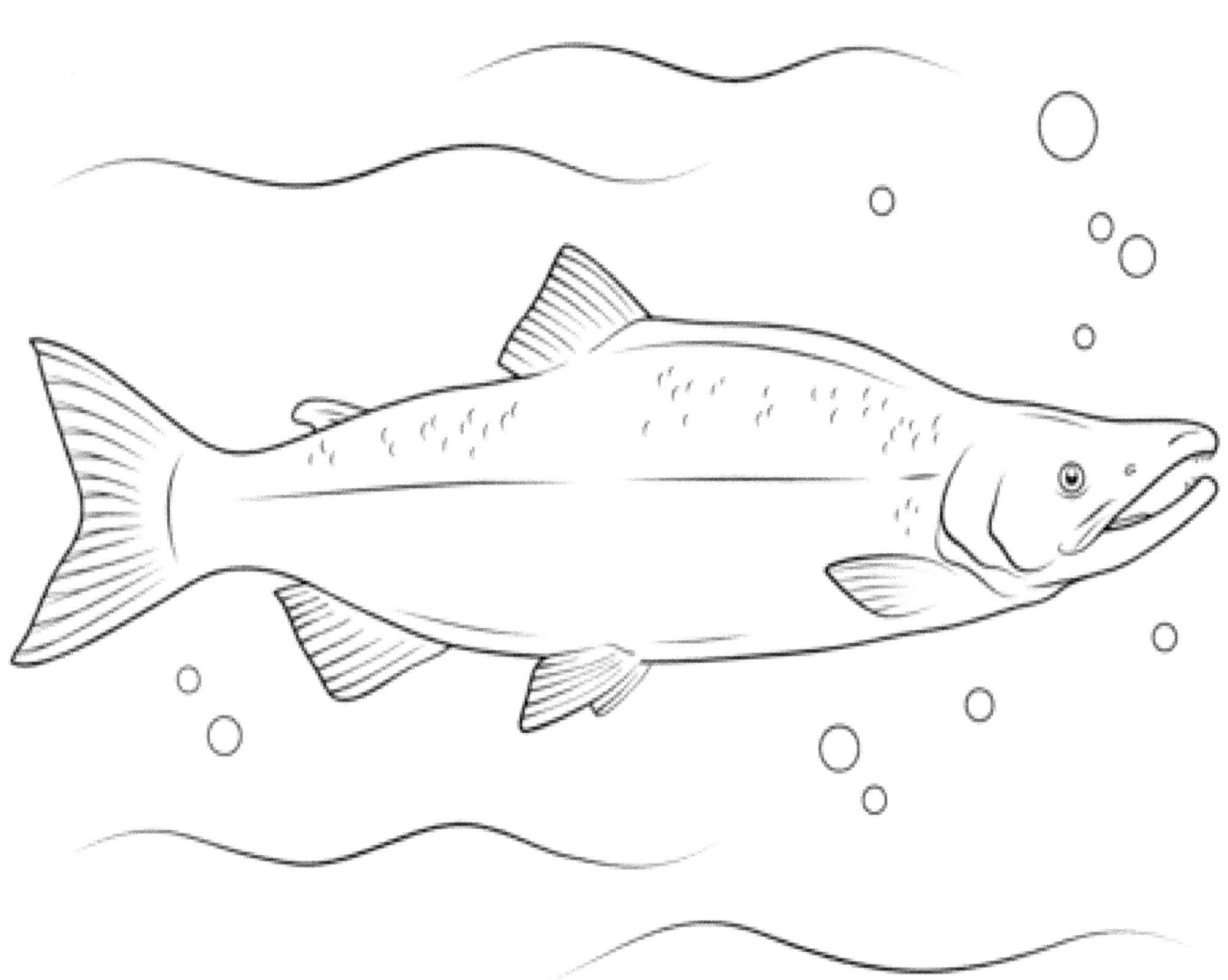

Sperm Whale

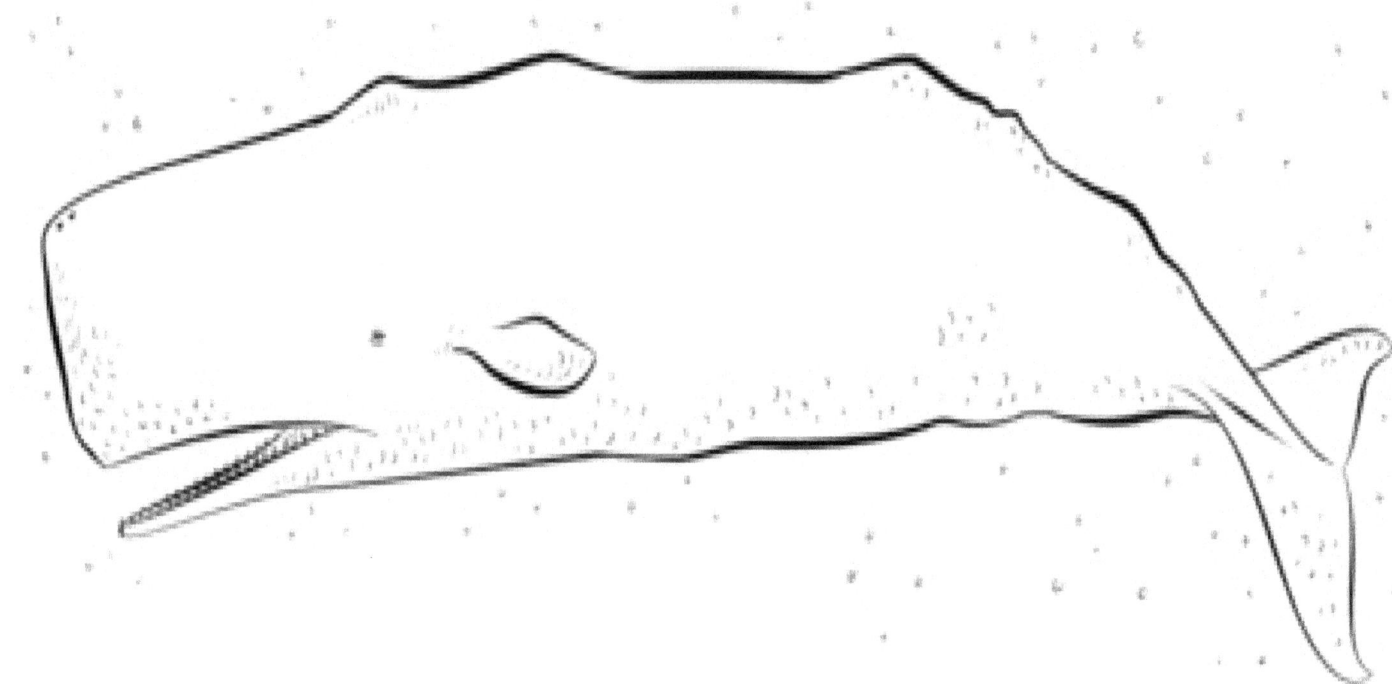

Starfish (Sea star)

Swordfish

Spiny lobster

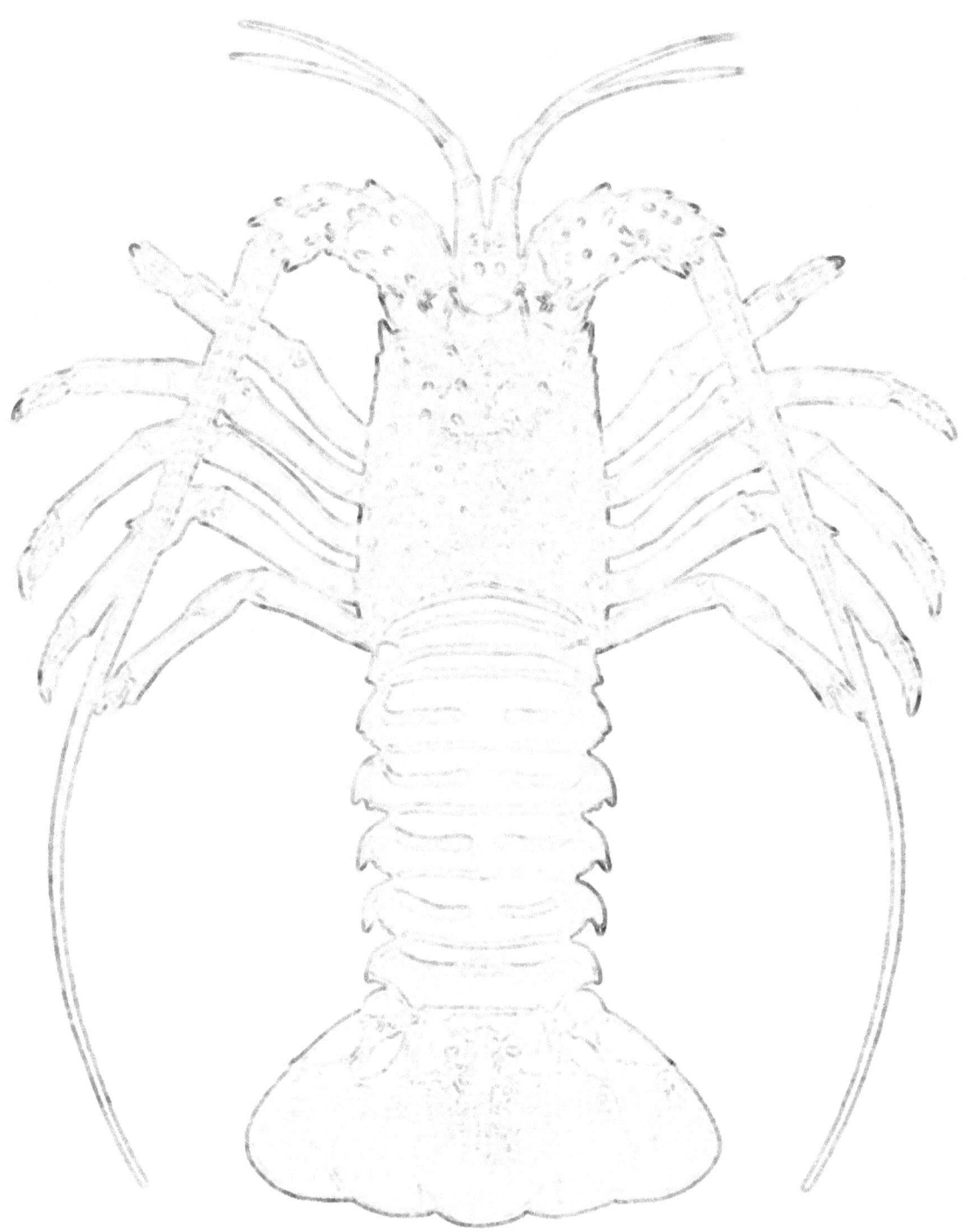

Sea Turtle

Tasmanian Giant Crab

Walrus

Yellow Tangs

Aquatic Fish

Sperm Whale

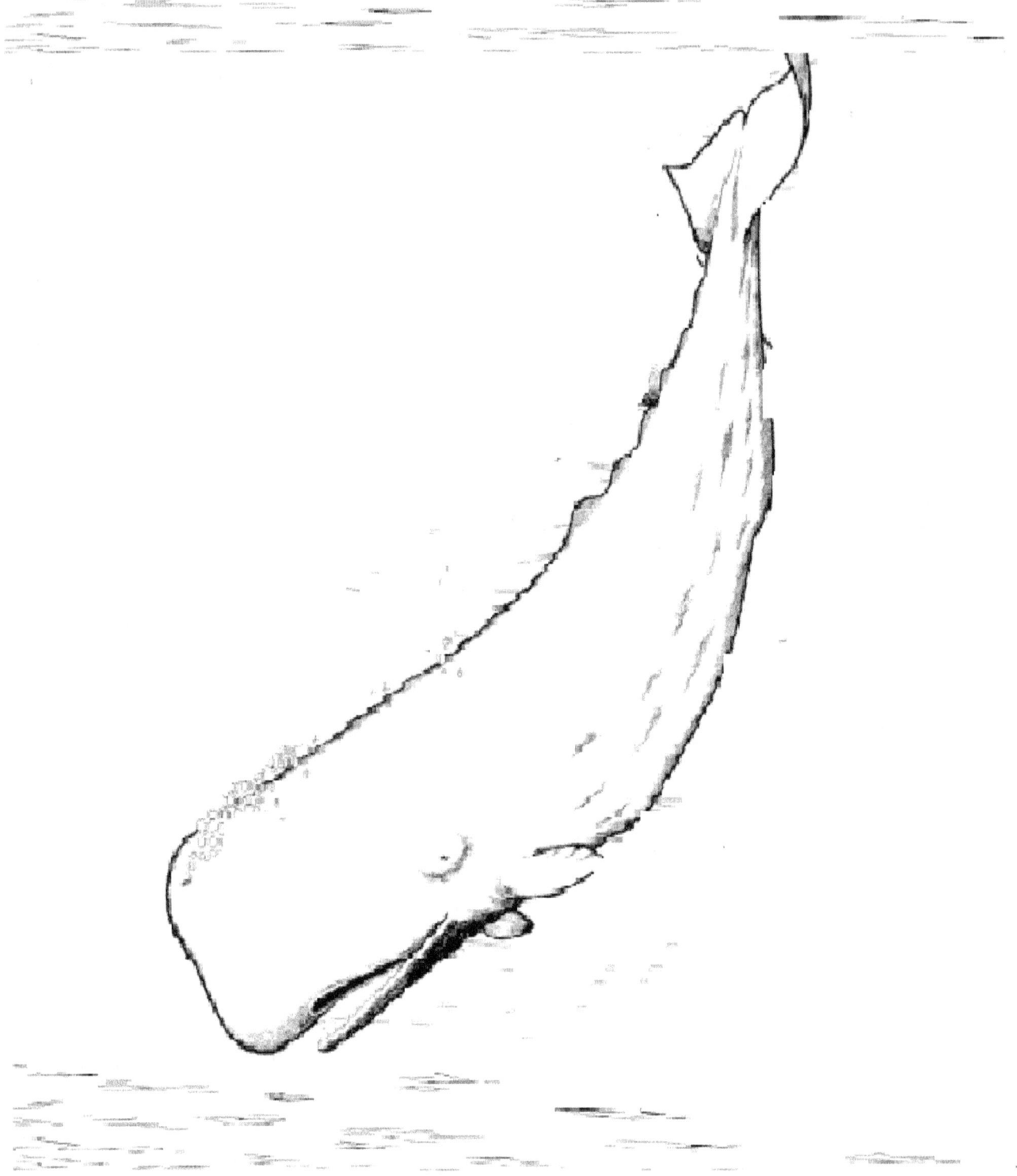

Giant Seal

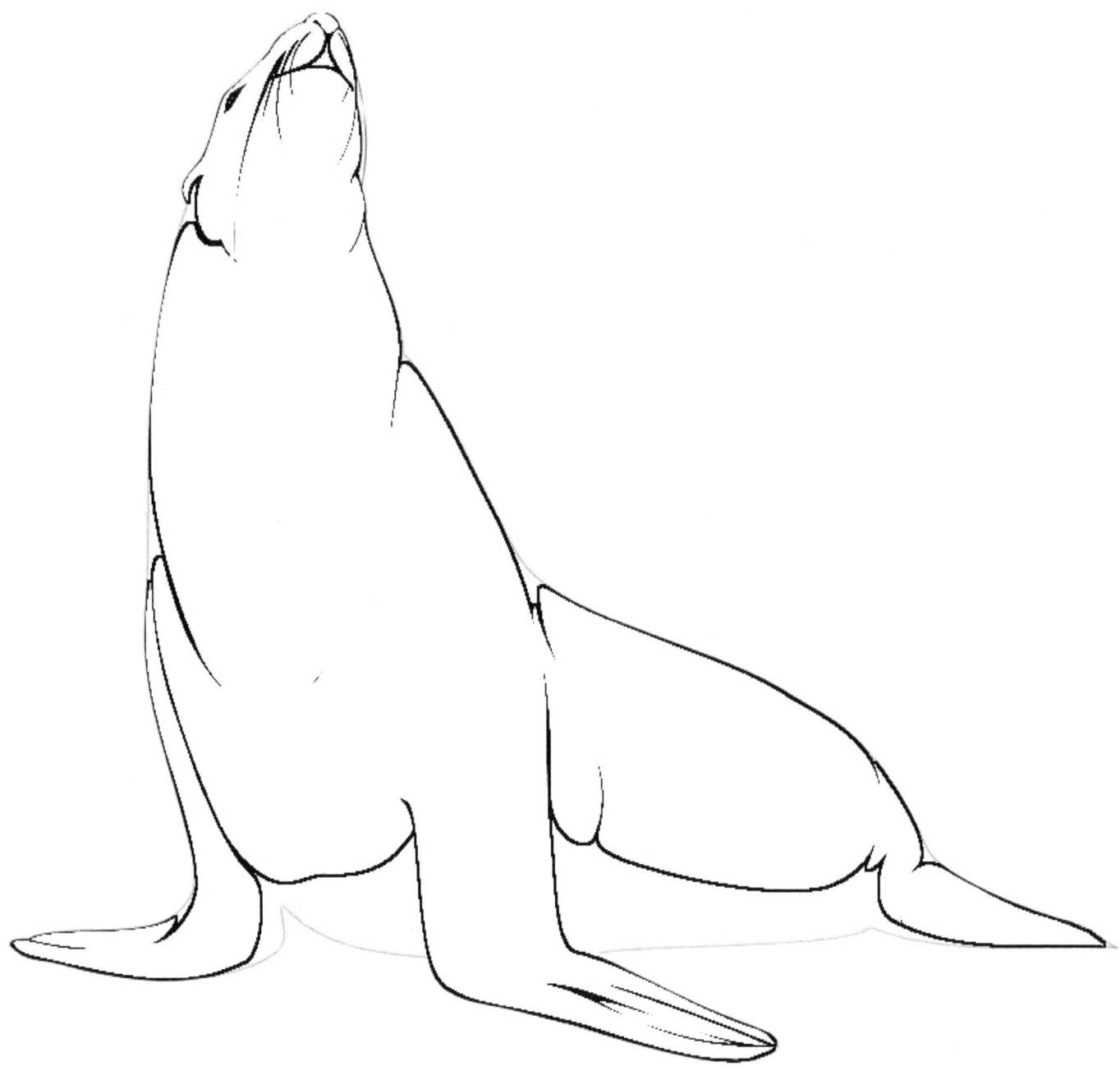

Velvet whalefish

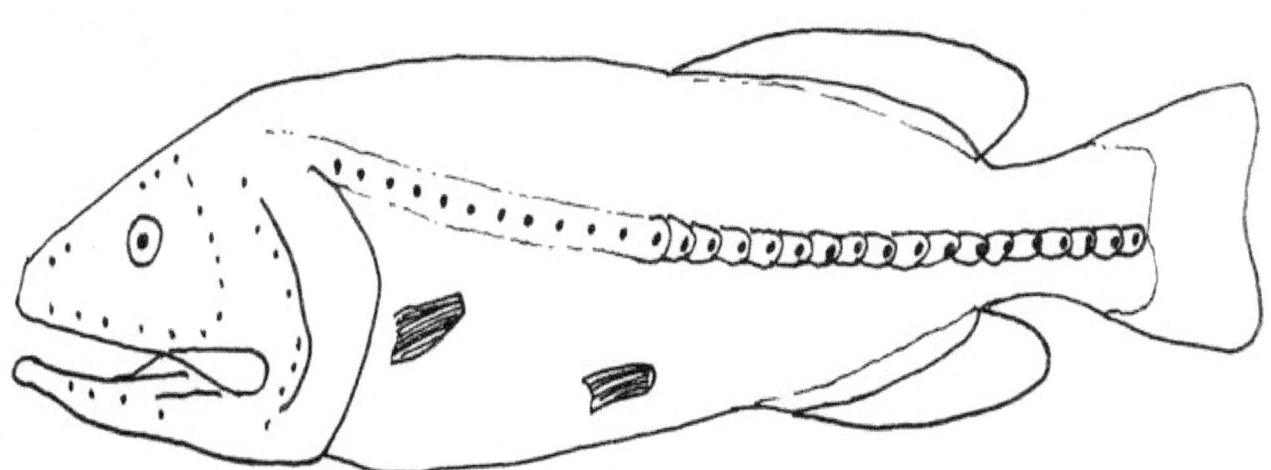

Shrimp

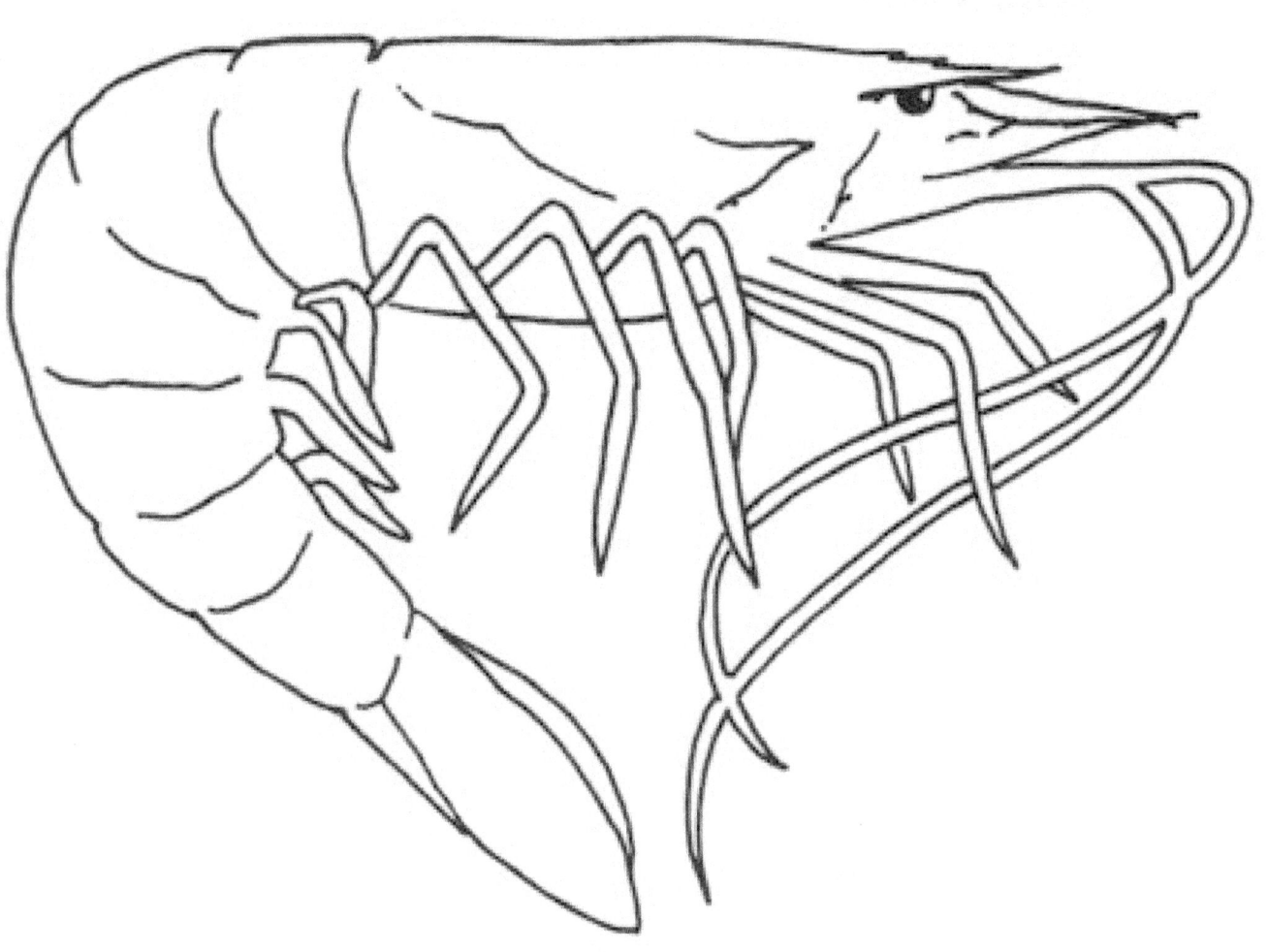

Bass

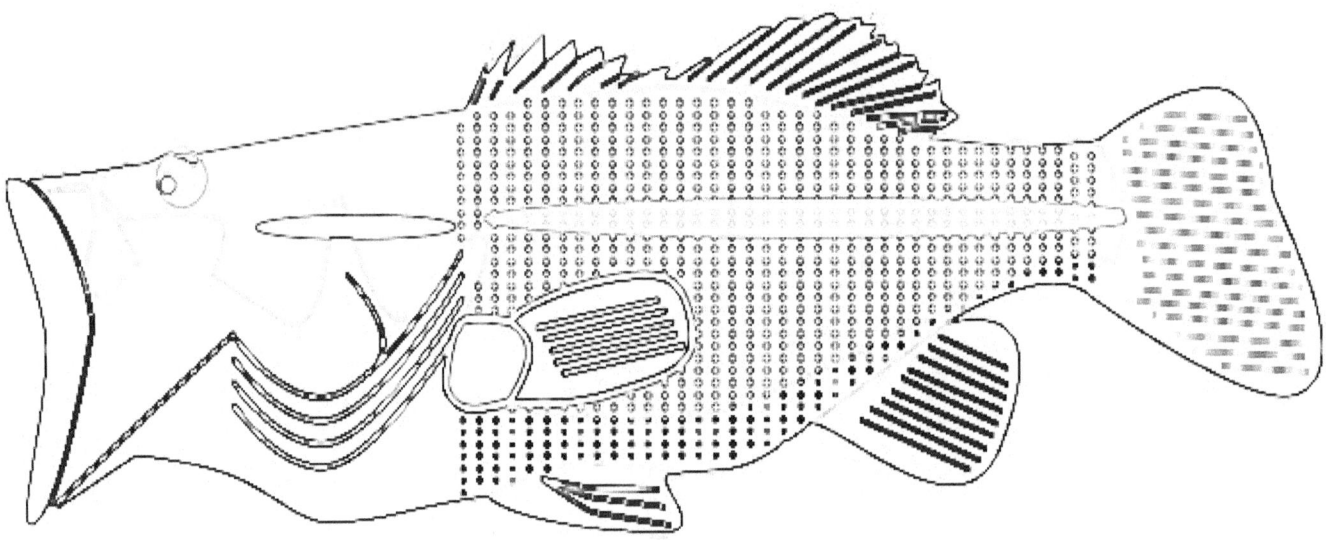

Bleu Fish

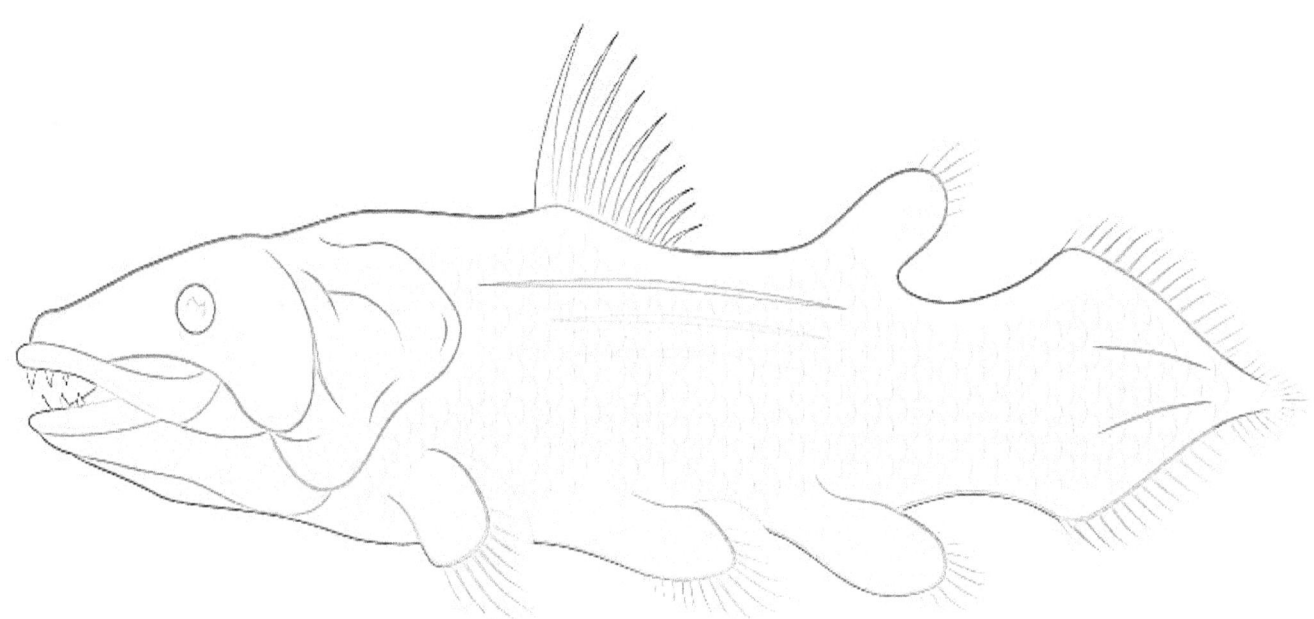

Common Cod

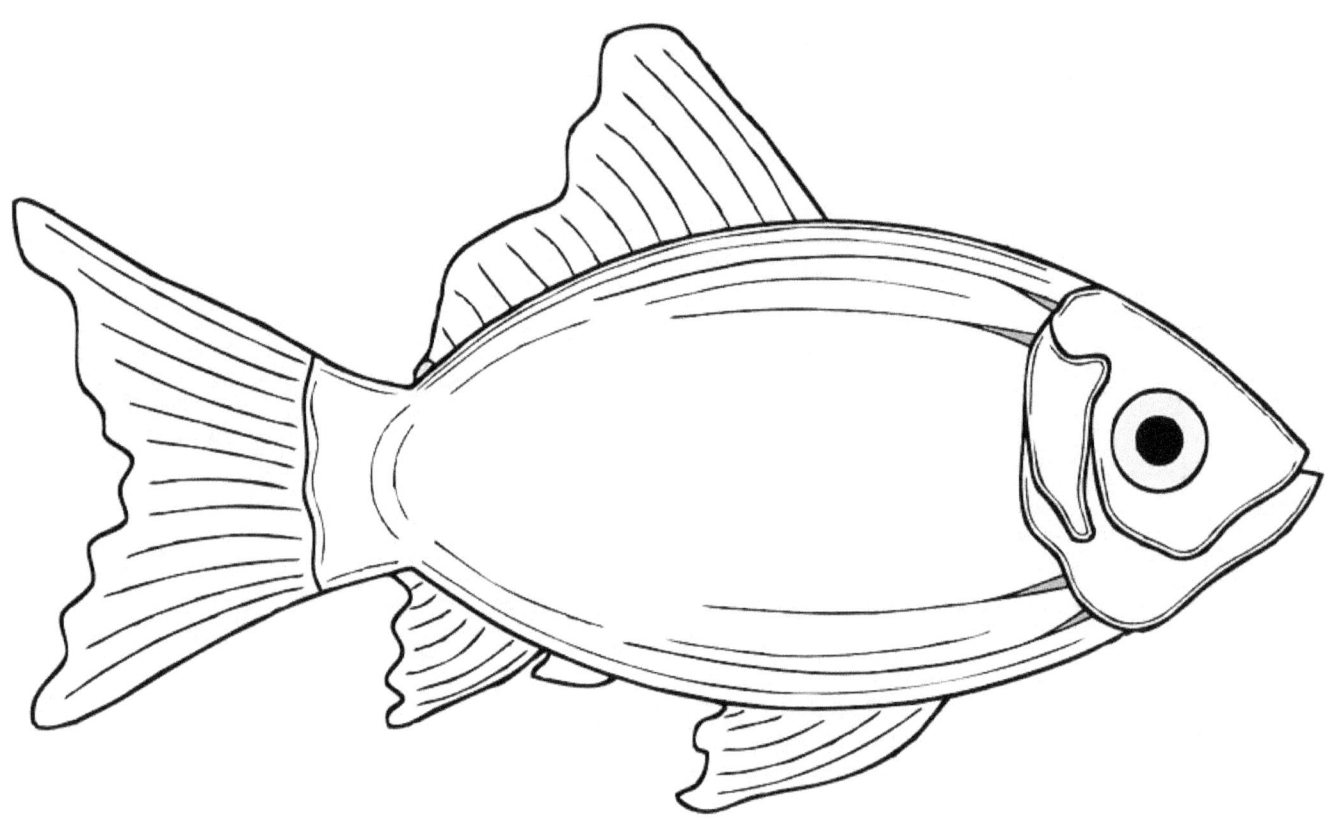

Common Carp

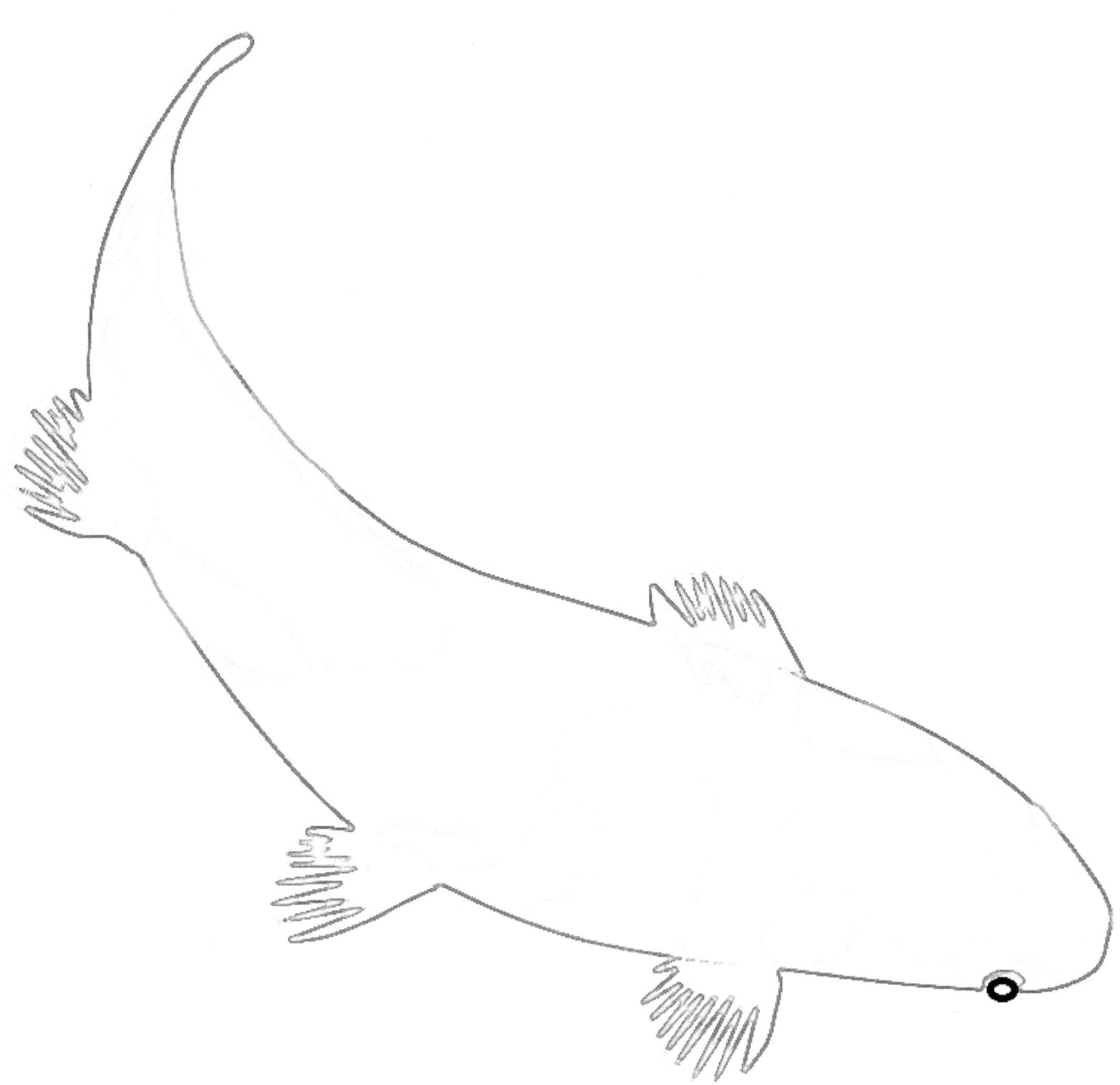

Deep sea fish

Dolphin

Octopus

Lantern fish

Perch

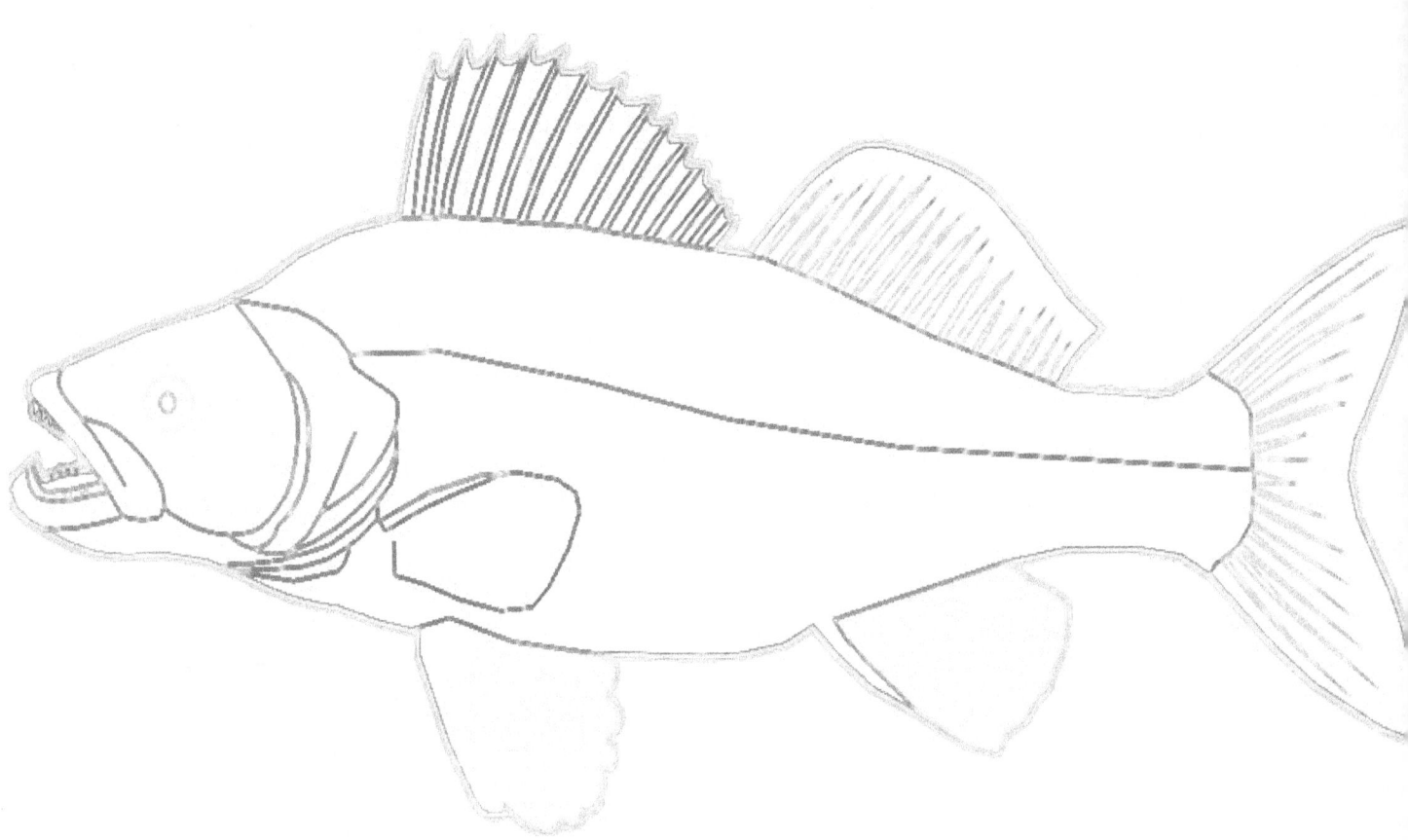

Whale

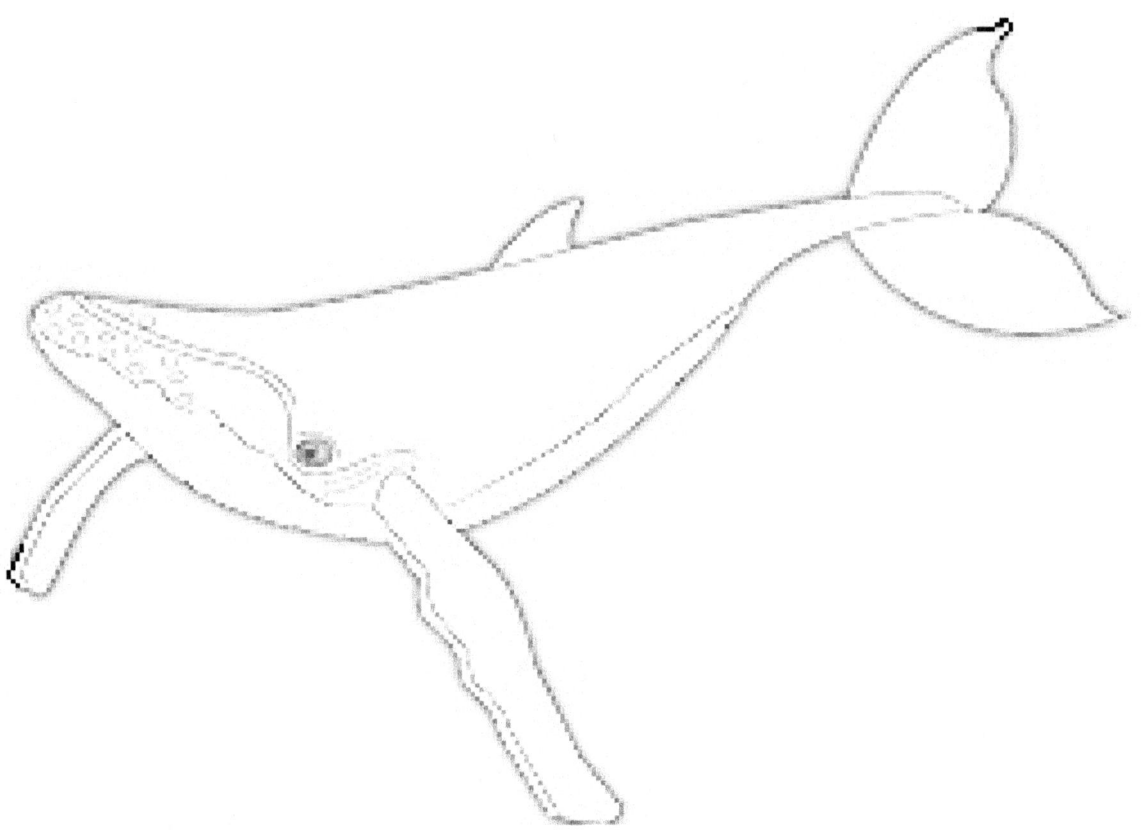

www.ingramcontent.com/pod-product-compliance
Lightning Source LLC
Chambersburg PA
CBHW080706190526

45169CB00006B/2257